Bibliographic information published by the German National Library:

The German National Library lists this publication in the National Bibliography; detailed bibliographic data are available on the Internet at http://dnb.dnb.de .

Imprint:

Copyright © 2017 GRIN Verlag, Open Publishing GmbH
Print and binding: Books on Demand GmbH, Norderstedt Germany
ISBN: 9783668455610

This book at GRIN:

http://www.grin.com/en/e-book/366122/systems-engineering-of-a-reusable-container-program-in-san-luis-obispo

Andrew Seaman

Systems Engineering of a Reusable Container Program in San Luis Obispo

GRIN Publishing

Systems Engineering of a Reusable Container Program in San Luis Obispo

By
Andrew Seaman
Cal Poly State University

ABSTRACT

San Luis Obispo is a progressive town and a leader in waste reduction striving to become a zero waste community. Currently, restaurants generate a large amount of waste from one time-use take out containers. Substituting one-time use containers with reusable containers will significantly reduce community waste. The following paper will perform a systems engineering analysis on creating a reusable container program in San Luis Obispo. Since the reusable container program is only in the design phase, forecasts will be made in the integration, production, and operations phases.

TABLE OF CONTENTS

Introduction.. 4

1. Needs Analysis.. 4

2. Concept Exploration... 6

3. Concept Definition.. 7

4. Decision Analysis and Support... 9

5. Advanced Development... 10

6. Engineering Design... 12

7. Integration and Evaluation.. 15

8. Production and Deployment.. 17

9. Operations and Support.. 18

Conclusion and Future Work.. 20

References... 21

INTRODUCTION

California Polytechnic State University has a newly developed reusable container program, headed by Dr. Tali Freed of the Industrial and Manufacturing Engineering department. The program is in the developmental stage and aims to secure an educational loan of up to $2 million dollars from the U.S. Department of Education. Several of Cal Poly's faculty from various departments as well as a large group of students have put time in the planning phase to create a reusable container program to help reduce waste in all of San Luis Obispo. The program revolves around take-out or togo containers from restaurants all over San Luis Obispo, Cal Poly included. The constant flow of students, travelers, and permanent residents creates a huge amount of container waste and these one-time containers can be eliminated. Already, several prototype reusable containers have been created and the program wants to implement new containers to stop plastic and paper food containers from going directly into landfills. Restaurants will serve food in a standardized container and customers can then drop containers off in various drop off bins around the community. After drop off, a service will clean the dirty dishes and provide participating restaurants with sanitized clean containers. Since no program like the one described currently exists, a grant will provide the necessary funding to demonstrate feasibility with wider adaptation expected in the future. The program ideology closely resembles reusable bags in grocery stores, and communities that ban plastic bags.

The following sections will discuss the concept development, engineering development, and post development phases of the system. Since many different engineering disciplines work together for a common goal an overall systems engineering plan will help bridge the lesion between different parties. The project does not have the necessary funding yet, therefore, engineering development and post development sections will be forecasts and estimates. These forecasts will add additional evidence to help secure the grant, which will eventually implement the program.

1.   NEEDS ANALYSIS

The San Luis Obispo reusable container program creates a brand new system. Therefore, focusing on how the system address human needs gauges marketability. Analyzing all stakeholders that are affected by a new system will form the driving factors for system application. Figure 1 shows a prioritization matrix of all stakeholders ranked by power and interest level concerning reusable togo containers. In addition, Figure 2 shows a prototype reusable container to help visualize the product.

Figure 1

<table>
<tr><td></td><td>Keep Satisified</td><td>Manage Closely</td></tr>
<tr><td>High</td><td>U.S. Department of Education</td><td>City of San Luis Obispo

Participating Restaurants</td></tr>
<tr><td>Power</td><td>Monitor
Service contractors
Container provider
Garbage company</td><td>Keep Informed
Students
Residents
Travelers</td></tr>
<tr><td>Low</td><td>Low</td><td>High</td></tr>
<tr><td></td><td colspan="2" style="text-align:center">Interest</td></tr>
</table>

Figure 2

The prioritization matrix above can be used to develop a detailed stakeholder analysis. Each party shown will in someway have an interest level as well power related to the program. The U.S department of education, City of San Luis Obispo, and participating restaurants hold the most power because they will regulate, fund, and require the program. Users like students, residents, and travelers will have high interest level, but ultimately rely on the regulatory stakeholders decisions. Contractors who will perform day to day execution tasks will benefit from the employment, but have little say in initial funding and planning. Next, Figure 3 below will take each stakeholder and rank level of support and influence strategy tactics to achieve or maintain support levels.

Figure 3

Stakeholders	Level of Support			Comments/Level of Support		Influence Strategy Tactics to achieve or maintain level of support
	Against	**Neutral**	**Supportive**	**Positive**	**Issues and Concerns**	
1) U.S Department of Education		X		Trial for future use	Expensive, cost reduction	Demonstrate benefits outweigh costs
2) City of San Luis Obispo			X	Progressive ideology, supports waste reduction	Cost, inconvenience, disruption of laws	Develop a detailed accurate plan to put reusable container program into place
3) Participating restaurants	X			Less expense if no cost on reusable containers	Hassle to change, concerned about expenses	Show restaurants that reusable containers will save them money

4) Service Contractors			X	New jobs, employment opportunity	None	Keep informed on number of new positions
5) Container Providers			X	Revenues and investment opportunities	None	Give sufficient lead time for orders
6) Garbage Companies	X			None	Less garbage, transition to new program jobs	Funnel garbage company jobs into reusable container jobs
7) Students, Residents, Travelers			X	Reduces waste	Inconvenience to drop containers off	Solve operation research problem to minimize inconvenience

Using the prioritization together with stakeholder analysis, areas of concern and benefit can easily be viewed. A detailed economic analysis will be necessary in the planning phase to satisfy not only concerned stakeholders such as garbage companies and restaurants, but also sources of funding such as the U.S Department of Education. These stakeholders will want concrete analysis on how much waste will be reduce, what the cost will be, and how the goal will be achieved. Demonstrating a positive benefit beyond reasonable doubt to each party will bring more groups to the strong positive side making transition easier.

Some helpful information on the current state system can be given to stakeholders to reduce concerns or move them to the positive support side. According to the US Environment Protection Agency, "the US generated about 229.2 million tons of municipal solid waste in 2001 and 32% was attributed to food packaging material"[1]. The millions of tons of food packaging waste informs stakeholders that food container waste is an important problem. Further contributing to the EPA statistic, college students tend to eat out more than the average population. College student act as catalyst to make San Luis Obispo a perfect town to test the program.

2.   CONCEPT EXPLORATION

Requirements analysis, concept synthesis, and feasibility experiments will all be discussed in the following section. Operational requirements analysis takes an engineering approach beyond needs analysis to justify spending. Figure 4 shows a list of operational and performance requirements pertaining to the system as a whole. Optimal drop off container locations, sanitation, ergonomics, economic costs and city requirements are several key factors will be addressed.. In order to understand the requirement regarding drop off bind, Figure 5 shows OZZI [2], a similar reusable container vending machine.

Figure 4

Operational Requirements	**Performance Requirements**
The drop off bins shall be conveniently located.	The containers shall have standard sizes.
The drop off bins shall have a pull down handle.	The City and restaurants shall provide customer with program instructions.
The drop off bins shall be easily understood.	People shall only be allowed 3 containers checked out at a time.
The program shall be ergonomic and simplistic.	The drop off bins shall be 6 feet high.
The containers shall be sanitized to a standard.	The dishwasher shall wash 500 dishes an hour.
The program shall be slowly implemented.	The program shall have simple (understandable) fees.
Users shall have little payment for the service.	

Figure 5

Using the list of operational and performance requirements along with case students a general system can start to form along with subsystems. Currently no county -wide system exists with drop off bins but, UC Berkeley has a reusable container system in place that covers 29% of the student population [3]. With an 80% return rate of containers and minimal incentives feasibility of a large system seems within reach. For the operational requirements the overall reusable container system is broken up into 3 categories: participating restaurants, drop off bin system, and cleaning/reuse of containers. Each category requires appropriate economics justification and acts as a system within itself. Participating restaurants will need to estimate their current expenses used for food containers each months and volume of containers used. Using data an estimated current cost per container can be developed for a two-year life cycle of a reusable containers. Furthermore, a cleaning fee per containers will need to be estimated for the drop off bin subsystem. Based off volume of containers processed per week a dollar value can be assigned to a container every time it is cleaned.

3.   CONCEPT DEFINITION

The concept definition phase will balance capability, operational life, and cost of the reusable container system. This section will perform an analysis of alternatives and system architecting. Since many different options are available for each subsystem this section will focus on 5 key components: containers, transportation, washing facilities, drop off bins, and payment plans. Each of these components have various options with no clear solution so evaluating multiple concept by performance, cost, schedule, and risk will result in an optimal solution. Figure 6 shows a table of potential options, favorable options are highlighted in green.

Reusable Container Program in San Luis Obispo - Formulation of Alternative Concepts			
System Component	**Concept 1**	**Concept 2**	**Concept 3**
Containers	Semi clear solid green plastic with compartments for food groups	Square 32 ounce black containers with a clear lid.	Opaque square solid black containers with compartments for food groups
Transportation	Several commercial size vans	One 14 foot box truck	
Washing Facilities	New washing facility	Use existing kitchens during closed hours	New washing facility in tandem with existing
Drop Off Bins	10-15 drop off bins	Drop off at participating restaurants only	Combination of drop off bins and restaurant drop off
Payment Plans	Users pay monthly fee	Restaurants pay monthly fee	City pays monthly fee

Figure 6

Concepts highlighted in green are selected based off relative goodness. Selections are based off cost, schedule and performance though, no concrete numbers have been selected yet. Once economic analysis and risk analysis are completed for washing facilities, drop off bins, and payment plans a more accurate decision can be made. Transportation needs to quickly and effectively more dirty and clean containers on a daily basis. Several catering style vans will be optimal for such tasks due to the large area covered. Washing facilities will depend on agreements between the City of San Luis Obispo and respective participating locations. Using existing kitchens would be easiest, but realistically using one kitchen will not be sufficient for all containers. Due to multiple optimal choices for reusable container style a multiple criteria decision analysis (MCDA) was complete in Figure 7. Three container styles were shown in the chart: semi clear green with compartments for food groups, square 32 oz black containers with a clear lid, and opaque square solid black containers with compartments for food groups [4]. These 3 choices were assigned numbers 1, 2, and 3 respectively.

Criteria	Cost	Privacy	Reliability	Ease of use	Total
Weights	6	7	7	5	
Container 1	6	4	7	6	143
Container 2	8	7	5	8	172
Container 3	6	8	6	6	164

Based off the MCDA analysis of container type the 32 oz black container with a clear lid would be the optimal choice followed closely by the opaque square solid black containers. However, since different restaurants serve different styles of food having multiple options may be beneficial. Using container 2 and container 3 together may satisfy customers the best way.

4.   DECISION ANALYSIS AND SUPPORT

Continuing on from the concept definition section the decision analysis section will attempt to reduce decision bias, begin to analyze data, and create models that help solve the overarching problem. First, potential decision biases will be identified to help mitigate incorrect thinking. Curse of knowledge and survivorship bias pose the highest risk when making decisions for this system. There is a huge amount of data on traffic volume, pedestrian volume, and waste in SLO and it will be difficult to sort through all available data which may be shown in a curse of knowledge bias. Furthermore, the survivorship bias needs to be addressed as this program will be successful from cost reduction and a reduction of overall waste. If waste output does not go down even with successful adaptation the program will not be.

Next, a system context diagram will be created to model each subsystem and see how the system functions as a whole. Figure 8 shows a system context diagram of the reusable container program with five subsystems, modeled as squares, surrounding it. The arrows help visualize what each specific subsystem gives and takes from the overall system. Each square is vital to make the system work, and if one square decides to leave the system the container program will most likely collapse. The system context diagram helps relate back to the stakeholder analysis and prove that every stakeholder needs a neutral or positive level of support.

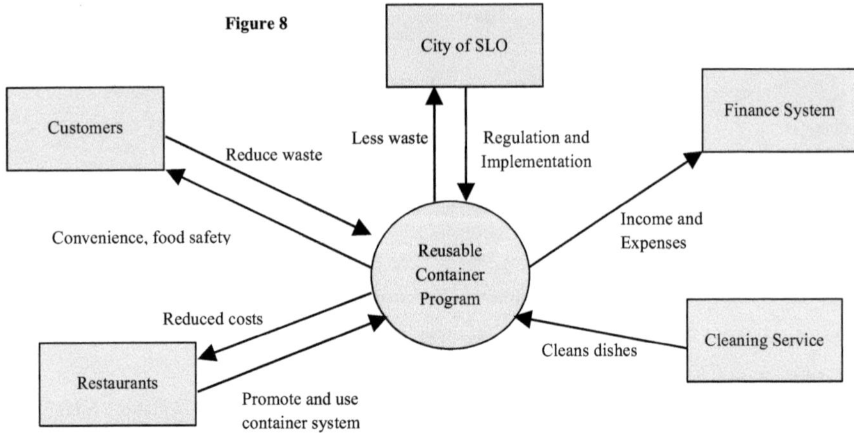

To conclude the concept development phase verification and validation testing must be done to determine if the right system is being build and if the system is being built correctly, with an overarching goal of reducing waste. According to rethinkrecycling one of the best ways to reduce waste is to, "bring reusable containers when shopping, avoid single-serve containers, and to be aware of double packaging"[5]. Since reusable containers helps reduce waste according to recommended waste reduction actions, the system appears initially verified . Furthermore, using metrics, such as number of one-time use containers used weekly for participating restaurants, can help determine if the system is being built correctly. If participating restaurants go from using 400 one-time use containers a week to zero a net gain of 400 less containers in the landfill can be concluded.

5.   ADVANCED DEVELOPMENT

The concept development stages have been proved in the previous sections. Now the engineering development stage will begin. This is the second system life cycle phase out of the three cycles: concept development, engineering development, and post development. This section will help identify risks in the system, reduce risks, and develop system design specifications. The largest risk with the container program is not getting funding and none of the functions can even be tested until then. In addition, delaying funding, or partial funding will result in an incomplete program. Deadlines need to be set and met regarding applying for grants, getting grant funding and making initial purchases. Beyond the risk of a lack of funding a Risk Matrix is shown in Figure 9. The risk matrix takes economic, schedule, technical, and programmatic risks then ranks them by probability and severity.

Severity		Highly Unlikely	Unlikely	Possible	Likely	Very Likely
	Extensive			Customers do not use program	No Funding	
	Major			No approval from City of SLO		
	Medium		Program doesn't reduce waste			
	Minor		Program is confusing	Homeless go through collection bins	Expenses are too high	
	No Impact					
		Highly Unlikely	Unlikely	Possible	Likely	Very Likely
				Probability		

Figure 9

Since there are numerous unknowns with a contemporary system and only a few smaller systems to reference the risk matrix shows relatively high risk with many items in the yellow zone and some in the red. The major two identified difficulties are not receiving funding and customers not using the program. Steps and laws can be made to force customers to start using the program. According to the National Conference of State Legislature [6] Los Angeles, San Francisco, Seattle, Chicago and Cambridge have banned plastic bags all together. In addition, California is the only state that banned plastic bags and requires reuse programs. The plastic bag program in California can be used as a model for the reusable container program. Setting laws into motion could prevent businesses from even offering one-time use containers and eliminate many risks in the program. The risk matrix helps show that San Luis Obispo is a formidable stakeholder in the project and could determine success. Already, Dr. Tali Freed has spoke with the mayor and the mayor is aware of the design phase of the program, which is an excellent initial step. Getting further legislative espousal and a law in tandem with funding would mitigate risk. Other risk areas that can be reduced are a confusing program as well as high expenses. Possible solutions for the confusion risk incorporate having directions on every drop off bin around the city as well as every restaurant. The drop off bins are sizeable and many people will see them visibly since the are placed in highly traveled areas. Once people realize what the purpose of the program is and how to use it adaptation will increase. Detection of failure will be relatively uncomplicated since data can be kept on how many containers are being washed per day. Similar to velocity of money a velocity of containers can be kept to determine how fast the containers move through the system as a whole.

6.   ENGINEERING DESIGN

Currently, the system is part way through the engineering design phase. In the next 10 weeks a final prototype will be competed on where to place the collection bins, what routes pickup and distribution will take, and a simulation model to mimic customers. Data has been gathered from three main sources: manual counting of people using one-time use containers in high-density location, online surveys, and accurate traffic volume data provided by the city. From literature review a report titled, "Pedestrian Count Method at Intersections: A Comparative Study" [7] discussed manually counting with counting sheets, clickers and video cameras. The report discussed ways to execute each method and found that manual counts with clickers underestimated error rates between 8-25%. This report proves that data taken with clickers is very precise and if anything will error on the side of caution. Figure 10 shows traffic data on all of San Luis Obispo.

Figure 10 [8]

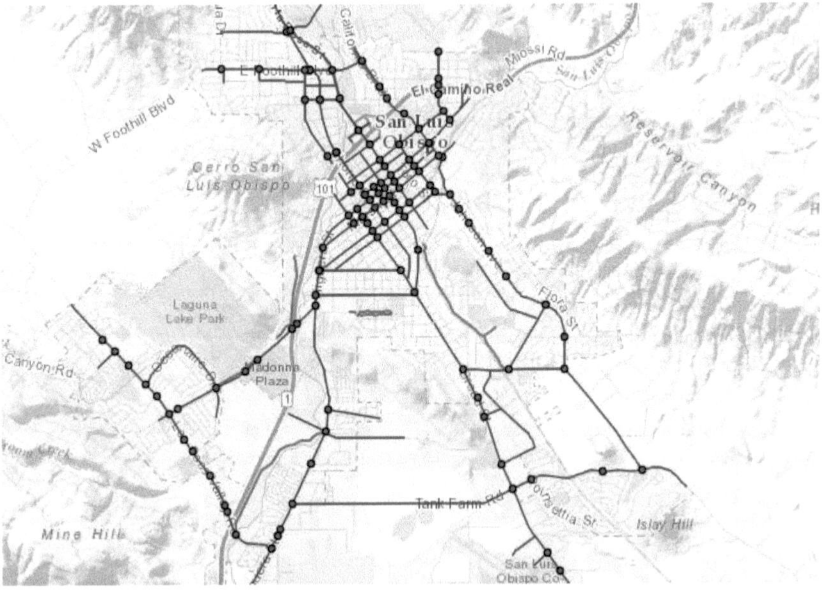

Each blue dot represents a stop light intersection and each red line represents a road that tracks data. When clicking on one of the components information is displayed that has daily traffic volume, pedestrian volume, turning volume, and pedestrian crossing volume. In addition, the data shows hourly volume so well traveled times of day can be observed. In total there are 113 stoplights with car and pedestrian data points. Figure 11 shows the top 15 intersections that have the highest car volume ranked from top to bottom. Figure 12 shows the top 15 intersections that have the highest pedestrian crossing volume.

Figure 11

Location	Car Volume
SANTA ROSA & FOOTHILL	21705
TANK FARM & BROAD	19802
BROAD & ORCUTT	17447
OLIVE & SANTA ROSA	17292
MADONNA & 101sb	16633
HIGHLAND & SANTA ROSA	16405
MADONNA & LOS OSOS VALLEY	16387
SANTA ROSA & MURRAY	16273
FROOM RCH & LOS OSOS VALLEY	15980
SOUTH & BROAD	15972
& LOS OSOS VALLEY	15493
INDUSTRIAL & BROAD	14954
HIGUERA & MADONNA	14677

Figure 12

Location	Pedestrian Volume
HIGUERA & MORRO	5229
HIGUERA & CHORRO	5183
MONTEREY & MORRO	4058
HIGUERA & BROAD	3779
MARSH & CHORRO	3041
HIGUERA & OSOS	2453
MARSH & MORRO	2284
MONTEREY & CHORRO	2135
MONTEREY & OSOS	2134
FOOTHILL & CALIFORNIA	1678
PALM & BROAD	1538
MARSH & BROAD	1257
HIGUERA & NIPOMO	1251

These intersections tables are a great starting point to put drop off collection bins. However, each one needs to be investigated in person to ensure that a spot is available for the bin to be places. The high car volume roads can be used for drive by drop off, similar to dropping of mail at the post off drive through. Next, the high volume pedestrian intersections can be targeted to position drop off bins for a standard drop off. In addition to this data, data from the hand counters can be used to select several areas where the city lacks data. Since the budget is currently uncertain a finalized list will be made ranking 1 to 20. If funding for 10 bins gets approved then the top 10 bin selections will be used, and if 15 bins get funding then the top 15. Based off the current ideology, bins will be one of two ways for customers to return their used containers. All participating restaurants should keep a place to take customers used dishes that can be straightforwardly assessable for the pickup cleaning service.

Next, a few assumptions will be made based off a system integration and new data provided. The assumed data shows that the current bins system was not adequate, restaurants were running out of containers, and customers were not returning containers and checking out many at a time. All three of these assumptions are likely scenarios in the future so addressing them now is beneficial. To implement the design changes an engineering change order (ECO) will be completed as shown in Figure 13.

Figure 13

Design Inputs	Design Output	Design Verification	Required Training
Additional Drop off Bins	Increase drop off bins size or add larger bin capacity.	Forecast if new bin capacity meets demand.	Cleaning service needs to meet new volume
More containers needed in system	Order more containers from supplier.	Volume analysis on restaurant container use.	None needed
Improved container tracking system	Track with credit card, coin system, or checkout fee.	Ease of use, plausibility, and ergonomics.	Restaurants and cleaning service

Each design input in figure 13 requires an engineering design, design verification and required training section. In this case adding an engineering design almost acts as adding an entire contemporary system to the process. Needing additional drop off bins requires more data analysis. Since the drop off bins costs around $5000 increasing capacity may not be effortless and require significant investment. Running another operations research solver simulation and adding in a constraint that represents bins already in the system may give ranked solutions of new spots for drop off bins. Next, a new routing procedure may need to be made for the cleaning service and the cleaning service must have excess capacity to meet the new routes. The simplest design input in the ECO is the additional containers needed in the system. Adding more containers will require an order from the supplier and some lead time before the containers arrive, currently there are no know supplier constraints.

The concluding input, container tracking, requires literature review and analysis of options. One paper discussing tracking titled, "Information quality attributes associated with RFID-derived benefits in the retail supply chain" [9] by Carmine Sellitto, Stephen Burgess, and Paul Hawking analyzed RFID tracking information. Their finding showed RFID benefits in timeliness, tracking, and reduced labor, which resulted in elevated profit. Attaching RFID chip to every container could link them to people who check them out, but may be a problematic and expensive solution. Another article conversing credit carding tracking methods by Minda Zetlin [10] shows that demographic data combined with purchase history can provide personal information about customers. The method, called data mining, can predict what customers may want to buy. Instead of using the tracking from credit cards to disclose personal history, the tracking method could be used to check out containers to people. If each containers

was connected to one database, any location could tell how many reusable containers someone has checked out, or

how many one has returned. Alternatively, a person could be charged a flat rate of 50 cents when they purchase a

container and receive 50 cents when they deposit one or bring it back to the store. Dr. Tali Freed also discussed a

method of giving each person a token that represents a container. When they deposit a container into a drop off bin

a token is dispensed, and can be used at participating stores. Based off the three sources stated above another

MCDA analysis may need to be created to find an optimal tracking system.

7.   INTEGRATION AND EVALUATION

The integration and evaluation stage comes from three main parts: integration, developmental system

testing, and operational test and evaluation. Below Figure 14 shows a plan for a system integration approach

assuming funding has been completed.

Figure 14

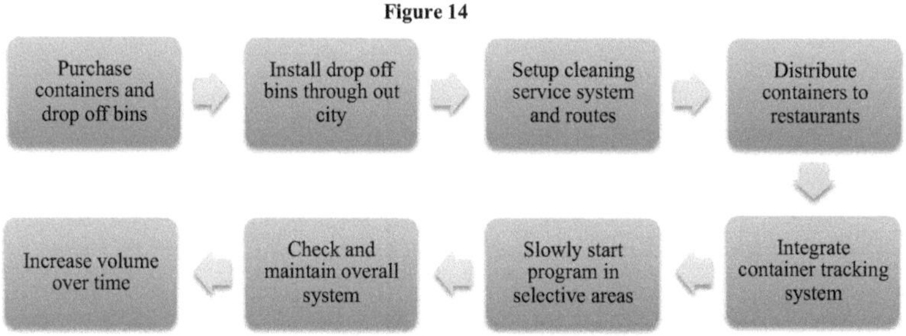

The process leads with purchasing containers and drop off bins square in the top left. This step assumes

that the 2 million dollar grant was passed, the City of San Luis Obispo has given full support to the program, and the

engineering design phase has been thoroughly completed. For the most part each step needs to be completed after

the previous step, but the distribution of containers and tracking system can be implemented at the same time. The

cleaning service can be outsourced to a third party or designed from scratch. Outsourcing may be simplifier since

hiring requires a human resources department, payroll, and additional complexity to the overall system. Integrating

the restaurants will require city employees or a task team to go door to door and explain the program goals. In the

beginning, a team consisting of the City of San Luis Obispo, key members of the project from Cal Poly, and

employees will need to take place to go through goals and objectives. Initial setup will require extra labor and effort to get each subsystems on board which will make a elevated startup cost. Therefore, a considerable portion of the funding will undoubtedly get used in the first few months, with an amount left over for maintenance.

Subsequent, developmental system testing, or verification, of the reusable container system will take place. This stage will test if the system reached technical requirements described in the concept development phase. With a product, testing phases are easier to outline. A engine may be tested for 400 hour, or a battery may be tested for capacity and power. Since, the reusable container program is unique, unique testing procedures will need to be made. In addition to developmental system testing, operational testing and evaluation will also need to be generated. Operational testing will validate if the system capture the customers needed and performs the intended function. Figure 15 shows several important subsystems with connected verification and validation tests.

Figure 15

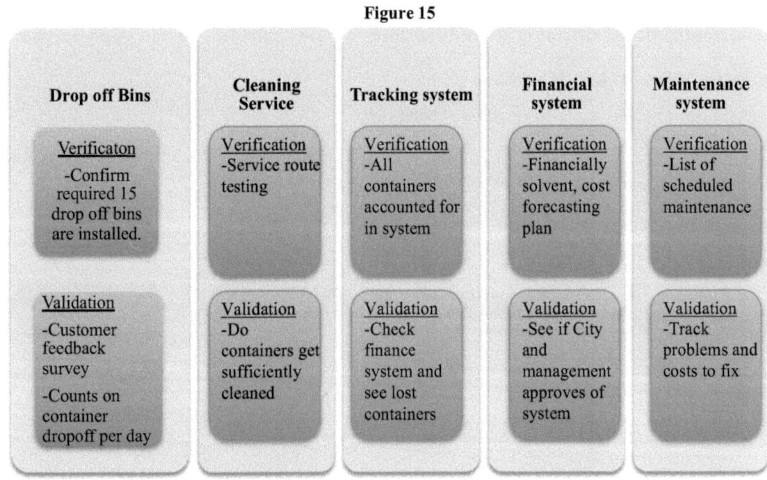

Testing was developed for drop off bins, cleaning service, tracking, finance, and maintenance. The top square's testing is design to see if the system meets requirements and the bottom square checks to meet customers' needs. In this case the customer tends to be the users of the container system and management. The drop off bin system can be verified by checking technical requirements like number of bins installed and having an inspector inspect quality of installation. Creating a customer feedback survey to see satisfaction of the drop off system and potential improvement will validate if the drop off bin system is the correct solution. The cleaning service can be verified by running a few dry runs of pickups and drop offs. Validation cleaning can be put up to a standard.

According to foodsafetymagazine, "cleaning is the complete removal of residues and soil from surfaces, leaving them visually clean so that subsequent disinfection will be effective" [11]. Inspecting cleaning before and after then comparing to food and safety standards will validate the system beyond a reasonable doubt. The tracking system and finance system can almost be tested together. A solid financial plan that is thoroughly audited will verify that grant money is used appropriately and doing weekly counts on bins in system and containers checked out will holding the tracking system accountable. While containers will most likely become lost in the system the rate needs to be at a reasonable level. Validation of the financial system can be check by providing audits to management, essentially checking that the customer is satisfied with the system in place. Finally, maintenance procedures can be written on paper, but they will need to be adjusted as more data becomes available. Scheduled maintenance for bin repair and container orders should be included in a plan. However, if no more containers need to be ordered additional expenses on excess container inventory shouldn't be followed just to abide by a maintenance system. Overall testing needs to ensure that the program satisfies the customers' needs, achieves its goal in reducing waste, and has a long-term financial plan.

8.   PRODUCTION AND DEPLOYMENT

The production and deployment phase is the second part in the post development system life cycle. This section will go over engineering for production, transition from development to production, production operations, and acquiring a production knowledge base. First, engineering production will be investigated using concurrent engineering processes to reduce initial costs. The final system is not a production system, but several main components do have a production line. The drop off bins, containers themselves, and cleaning service materials all have production lines. Concurrent engineering will help to reduce complexity, environmental impact, improve job design, and improve maintainability of each design.

First, the containers themselves can be standardized to save on costs. The existing design has three compartments for food and are semi clear. Figure 2 in the needs analysis shows the initial container design, which costs between $3-5 each and is semi clear so anyone can see the owners' food. Figure 16 [12,13] shows the other design concepts, which reduces complexity and standardization of components. In comparison, the alternative containers cost only $1-2 or potentially less since there is only one square spot for the food to go. Making deals with manufacturing including tax breaks due to environmental impact could reduce costs even further.

Figure 16

 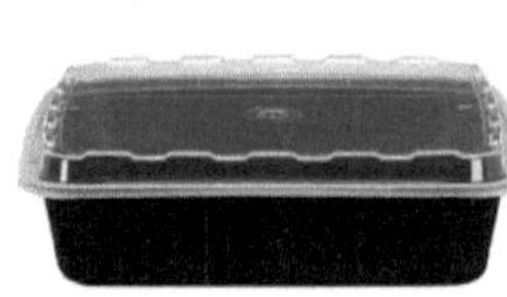 

Reducing costs by $4 a container through standardization could also lead to easier cleaning methods, resulting in reduced serviceability costs. If 10000 containers are in the system, simply switching designs could save nearly $50000 annually which could convince others to adapt the system. Furthermore, the drop off bins are estimated at $5000-$10000 a piece. Having mechanical engineering and manufacturing engineering students at Cal Poly redesign the drop off containers for ease of manufacturability could cut the costs to 10% of the price. Currently, the plan goes to known solutions with complex parts looking for large amounts of grant money. Once engineering teams begin working together for production of the key products instead of management simply trying to push the program, costs will go down.

9.   OPERATIONS AND SUPPORT

The final stage in the systems engineering process is the operations and support section. This section will discuss operational performance, regulatory issues, and disposal. Operational performance will take place right after the initial system has been completed. This might be after a few restaurants have containers and 5-8 drop off bins have been installed while the cleaning service has just begun their routing procedures. Figure 17 shows examples of upgrade periods where the container program will increase scheduled containers in system and drop off bins. These upgrade points will also include adding more restaurants into the system as well. The time period could be a constant 3 months or when milestones are reached in funding or performance. If the financial system confirms accountability with a certain time period and each testing solution described in the validation and verification testing passes then the next time period goal can be attempted. Also, while the program advances new technology should be investigated. According to a scientific publication by the University of Novi Pazar, "thanks to dropping costs RFID tags are likely to proliferate into the billion in the next several years – and eventually the trillions" [14]. Initially, a manual coin method may be the best option for tracking the container but checking in on RFID in 2-10 years may

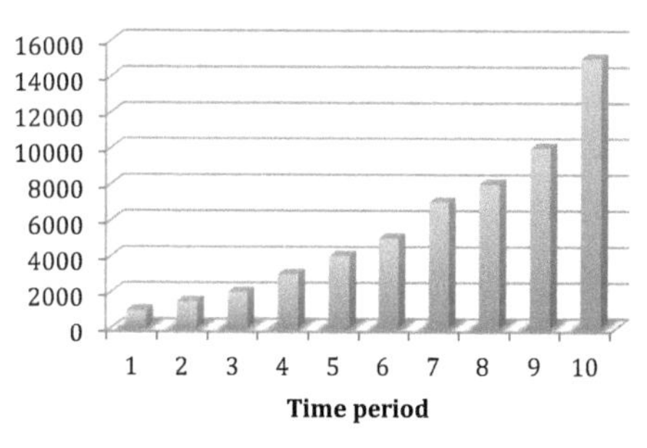

Figure 17

conclude that expenses have dropped significantly. At a certain price points it will be worth it to add RFID tracking chips. RFID tracking would be considered a major system upgrade. A entirely new dynamic database would need to be programming, which would require significant funding.

Another issue with the above graphs relates to facilities and personnel limitations. Each washing facility will have a capacity, and at a certain point they will not be able to handle any more containers in a day. The phase in solution helps, but additional facilities and employees may be needed to keep up with demand. A management system may need to be added to handle information, complaint, recommendations, and problems. Adding a full time manager into a city building, specifically for the reusable container system, may be necessary. Members of the Cal Poly team planning and setting up the program have other responsibilities and cannot commit to day-to-day operations. Luckily, the program currently runs on a volunteer base so the entire planning and funding currently get hundreds of hours of free expert work. Furthermore, additional senior projects could perform operations process and improvement calculations to improve the system and teach real world skills. Faculty and students alike realize the important of environmental impact and collective continued support will ensure smooth operations.

CONCLUSION AND FUTURE WORK

To meet sufficient demand for an entire county many container drop off bins, restaurants, and washing stations will need to exist to bring a reusable container program to San Luis Obispo. Already students and teachers have completed the initial planning phase for the system, but a more detailed economic analysis needs to be completed to apply for a grant. The above report outlines the concept development, engineering development, and forecast of post development for the system as a whole. Findings show that costs in the planning phase can be significantly reduce with improved standardization and simplicity. The greatest risk comes with funding and the each system will need to be scalable for future growth. Overall, the reusable container program will significantly reduce waste in San Luis Obispo.

REFERENCES

[1] US Environmental Protection Agency, Office of Solid Waste and Emergency Response. (2001). *Municipal solid waste in the united states: 2001 facts and figures*. Retrieved from website: https://www.epa.gov/smm/advancing-sustainable-materials-management-facts-and-figures

[2] "OZZI IN THE NEWS." *Welcome to OZZI*. N.p., n.d. Web. 08 Feb. 2017.

[3] Harnoto, Monica F. "A Comparative Life Cycle Assessment of Compostable and Reusable Takeout Clamshells at the University of California, Berkeley." 2013.

[4] Co., Ace Mart Restaurant Supply. "Ace Mart Restaurant Supply." *Ace Mart Restaurant Supply*. N.p., n.d. Web. 08 Feb. 2017

[5] "Top 10 Ways to Reduce Waste." *Rethink Recycling*. N.p., n.d. Web. 15 Mar. 2017.

[6] "Fees, Taxes and Bans | Recycling and Reuse." *STATE PLASTIC AND PAPER BAG LEGISLATION*. N.p., 11 Nov. 2016. Web.

[7] Diogenes, Mara Chagas, Ryan Greene-Roesel, Lindsay S. Arnold, and David R. Ragland. "Pedestrian Counting Methods at Intersections: A Comparative Study." *EScholarship*. N.p., 04 Apr. 2007. Web. 15 Mar. 2017.

[8] "SLO City Map." *Slocity.maps.arcgis.com*. N.p., n.d. Web. 15 Mar. 2017.

[9] Sellitto, Carmine, Stephen Burgess, and Paul Hawking. "International Journal of Retail & Distribution Management." *Emeraldinsight*. N.p., 1990. Web. 15 Mar. 2017.

[10] Zetlin, Minda. [11] "Cleaning and Disinfection: Improving Food Safety and Operational Efficiency in Food Processing." *Food Safety Magazine*. Diversity Care, n.d. Web. 16 Mar. 2017.

[12] "404." *Alibaba Manufacturer Directory - Suppliers, Manufacturers, Exporters & Importers*. N.p., n.d. Web. 16 Mar. 2017.

[13] "CuBE 48 Oz Black Bottom Square Container With Lid 100/Case." *ShopAtDean*. N.p., n.d. Web. 16 Mar. 2017.

[14] "RFID: Past, Present, Future." University of Novi Pazar, 2012. Web.

"The Latest Privacy Invasion: Retailer Tracking." *CreditCards.com*. Creditcards.com, 02 Feb. 2016. Web. 15 Mar. 2017.

YOUR KNOWLEDGE HAS VALUE

- We will publish your bachelor's and master's thesis, essays and papers

- Your own eBook and book - sold worldwide in all relevant shops

- Earn money with each sale

Upload your text at www.GRIN.com and publish for free